W.A.R. Stories!

Wisdom Acquired Randomly

BOB VÁSQUEZ

W.A.R Stories! Randomly Acquired Wisdom

Published in the U.S.A by Wajbook Press, San Antonio, TX.

Paperback Version: ISBN 978-1-7357466-9-2
Kindle Version: ISBN 979-8-9867720-0-4
Library of Congress Control Number: 2022944711

For information about special discounts for bulk purchases or speaking opportunities, please contact the author at bobvasquez@bobvasquez.com. The views expressed are those of the author.

CONTENTS

Wisdom is the merging of experience and knowledge.

Stan Strycharz

Wisdom is your perspective on life, your sense of balance, your understanding of how the various parts and principles apply and relate to each other.

Stephen R. Covey

Wisdom surrounds us. It is seldom hidden but often overlooked.

Noah benShea

Introduction

I'm a Wisdom Seeker. In my quest for wisdom, I've discovered that it's most often acquired randomly… through experience and insight. Wisdom Insights often occur serendipitously. You never know when you're going to learn something.

These stories, which are all true and happened randomly, have helped me become wiser, not necessarily wise, but wiser. They're intended to instill thoughts in you that might empower you to think deeply about what surrounds you daily and how you can make your world better. I'm hoping that they'll resonate with you and that you'll relate to them. I hope they help you become wiser, as well.

bob vásquez!

About the author

Bob Vásquez served at the United States Air Force Academy from 2002 to 2022 as a curriculum developer and as a course director for the freshman seminar offered through the Academy's Center for Character Development. He's an adjunct instructor at the University of Colorado at Colorado Springs.

Bob served in the United States Air Force for more almost 31 years before retiring in October of 2002. Bob invested 24 of those years serving with Air Force bands throughout the world. He also served as Commandant of the Noncommissioned Officer Academy at March Air Force Base, California; Senior Enlisted Advisor to the Commander of the 92nd Air Refueling Wing at Fairchild Air Force Base, Washington; Deputy Director of the Family Support Center at Ramstein Air Base, Germany; and Superintendent for the 86th Support Group at Ramstein.

He's the author of many books, the titles of which can be found at the end of the book.

A wisdom seeker, igniter, storyteller, musician, public speaker, life coach, and mentor, Bob considers his greatest accomplishments the successful raising of his daughters, Tesa and Elyse, six grandchildren, two sons-in-law, and growing closer to Debbie, his lovely bride of more than 44 wonderful and fulfilling years.

Connect with bob at
bobvasquez@bobvasquez.com.

Love unconditionally…

We'd JUST arrived in Germany. No kidding, day two. One of my new teammates, Susan, whom I'd never met before, invited Deb, our girls, and me, to her home for dinner. We accepted.

When we arrived, Joe, Susan's husband, was in the back yard, cooking. We all went outside to join him, introduced everyone, and got to chatting. Shortly afterward, Susan and Joe's son, Jeremiah, came out to join us. What an outgoing young man! He was about eighteen years old. Couldn't stop smiling, which was contagious, so we couldn't either. Didn't need a reason, other than Jeremiah was smiling.

Dinner was very nice. Joe, Susan, and Jeremiah were very gracious. As we walked out of their house on our way to our car, we said so long to everyone as Jeremiah came up to me, gave me a huge bear hug, which I can still feel at this moment, and said, "I love you, Chief." FREAKED ME OUT! This young man doesn't know me, how can he love me? Because love is unconditional. If you need a reason to love, that's not real love. BTW, Jeremiah has Down Syndrome.

Not that it matters, really, other than we sometimes think that people who are different than most of us are "handicapped" somehow. Ironically, Jeremiah may have Down Syndrome, but there's nothing DOWN about him. As I said before, his smile was contagious. He's one of the most UP people I've ever met.

That evening was the first and only time I ever saw Jeremiah. I stay in touch with his parents on Facebook. He's doing well. His dad says he IS unconditional love. I'm confident he is.

INSIGHT: The lesson here is this: love is unconditional giving. When you give, you get. When you love, you're not concerned about getting...it just happens. That hug Jeremiah gave me is so embedded in my heart that it will remain for as long as I live. The impact of that hug is that since then, I've hugged thousands of people with no intent other than for them to know I love them. Give unconditionally.... NOW!

My most interesting talk…

I'm sitting in my office, minding my own business, when my boss walks in, almost in a frenzy…. "Chief, are you available tomorrow at noon?" he asks.

"I am. What do you need?" I reply.

"I'm scheduled to speak to a group downtown at noon tomorrow, but something has come up that I can't make it. Will you go do the gig for me?" asks my boss.

"Surely," I answer. "Who will I be speaking to and about what?"

"It's a group of vets. You can talk about anything you'd like."

"Copy THAT, Sir! Can do easy!" I tell him. I'm a vet myself. I have plenty of war stories that they'll be able to relate to and I may even have a good message for them. This will be a piece of cake. I'm even excited about it.

Tomorrow comes and I get to the venue to speak. I do what I always do whenever I speak to a group. I walk around shaking hands, getting to know my audience so that I can relate to them in my talk. As I do so, I begin to notice that all of these people are really young to be vets. As I've grown more mature, not older, I've noticed that pretty much everyone seems young, or younger, to me, so I pay it no mind. Until I'm about to be introduced....

The MC starts out by welcoming everyone to the annual meeting of VETERINARIANS! Not VETERANS, but veterinarians! THAT'S why they're so young! Luckily, I'm fairly experienced at speaking to large audiences and I have a lot of stories in my arsenal, although I have to change the characters a bit. I get through it. I do well enough that they give me a standing ovation after laughing and almost crying when they were supposed to. My back-up topic, by the way, is always character and leadership. Everyone does that, or tries to, and I've been teaching in that area for decades. Being a storyteller helps. People appreciate and learn more from stories than from facts.

INSIGHT: The lesson here is simple. Communication is key to success. But it's never one-sided. Don't just ask the right questions, make sure you understand the answers. And when you answer someone's question, make sure it's in THEIR language, not yours. You may know exactly what you thought you said, but it may not be what the listener heard and understood. Veterinarians are vets, too!

Those wellington boots...

I had a cousin who wore Wellington boots. You know, black smooth-toed boots that were popular in the late 60's. Like Andy Griffith wore on The Andy Griffith Show. His dad wore them, too. I loved those boots and envied that they wore them. I don't recall if they were expensive, but I do remember that my dad couldn't afford to buy me a pair. They, to me, were a sign of prestige, maybe because no one else wore them. They were classy, a symbol of success.

About five years ago, Deb and I were window shopping in the Dillard's shoe department. Yeah, that makes no sense, window shopping in the shoe department of a clothing store. But they were having a big sale! We purchased some shoes for Deb, and feeling sorry for me, we went downstairs to look at men's shoes.

Just as we got to the men's shoes department, right at the front, displayed by themselves, was a pair of Wellington boots! I immediately had flashbacks to the days when I envied my cousin's boots. That feeling I had when I saw him wearing his boots overtook me, but only for a short while. Things were different this day!

You see, one of the most important lessons I learned from my dad, and I learned it by his example, was what we then called the work ethic. He was a hard worker, which I became, as I learned by watching him. He taught me that, first, it was the right way to be, and secondly, that if you work hard, you can succeed and get whatever you want....

On this day, I realized that all my hard work, all of the years that I'd practiced what my dad taught me, was about to pay off. I asked the shoe guy if he had my size in those Wellington boots. He went back to his storage room and returned with a pair. They fit perfectly! I asked him to put them in a bag as I paid for them. Cash! I, now, had the means to afford those boots! It'd taken several decades, but this was the day that I felt like a success.

I still have those boots. I don't wear them as much as I thought I would. But when I see them in the closet, it reminds me of my dad and how hard he worked to provide for his family and that I should do the same. I still do. I can't help it anymore. It's part of my identity.

INSIGHT: I'm not sure if we still even talk about having a work ethic. But I can tell you, it still exists. If you want something, the only way you'll ever get it is by working for it. Do the work!

Never messed up my shine…

Having just arrived at this particular assignment, I'm sitting at my desk, minding my own business, for once, when, all of a sudden, this Master Sergeant, whom I'd never met before, walks into my office, slams the door shut, and starts yelling at me. I'm totally taken by surprise. I try to listen, but I'm not used to strangers yelling at me, after all I'm a Chief! As she continues, I get frustrated pretty quickly and, literally, have to stand up, slam my hands on my desk, and shout above her to "Stand down!" She does, with the look of "Did I just do that?" on her face. I reminder her of who I am, and the relationship we should be expressing at the moment, which should be one of mutual respect, and that she's violating that standard. She apologizes, sincerely. I can see it in her face. As she calms down, she relates her frustration to me. We have a mutually respectful conversation after that, and she eventually leaves my office on good terms.

When I left that unit, that same Master Sergeant came by on my last duty day, closed the door gently, and in a very sincere and respectful way told me, “Chief, I just want to tell you that we had many discussions the past few years, and you sometimes stepped on my toes, but you never messed up my shine.”

INSIGHT: I don’t recall a LOT of disagreements with her, but I DO recall practicing what Dr Stephen Covey taught me, and that’s that we can agree to disagree agreeably. We don’t have to agree with each other to respect each other.

Not being poor…

I was invited to speak to a group of high-schoolers who were meeting at New Mexico Military Institute for a growth experience when one of them was asked to share his perspective on life-so-far.

Ever Vásquez stood in front of the group and talked about what it was like growing up Chicano and poor. He had had some profound experiences and had already, even though maybe only fifteen years old, matured beyond his years. His main theme dealt with being poor. He was, or had been, by any definition, except mine.

I followed Ever and followed up on his theme. He and his family were now better off financially, which was a blessing, so I was able to ask him, with the intent of providing some deep thought, "Does your family love you?" His reply was, "Yes, they do." "Then you're not really poor," I suggested.

A hush came over the crowd. Most, probably all, of the kids in the audience had been, or still were, financially poor, but each of them acknowledged that their parents, siblings, and all of their respective family members loved them. And they acknowledged that that blessing kept them from being really poor.

I remember that moment as if it was occurring as I write. A simple question gave each of those youngsters hope just by adjusting their perspective.

INSIGHT: I pray constantly for those youngsters, and all youngsters, that no one has to go without life's necessities and that everyone get much of what they wish for. But what they, and all of us, need most of all is love. And guess what? We can give that to others. It costs little if anything. Love someone today. Even a stranger. Show them compassion. Help them to not feel poor.

Don't just do it. Do it right...

You've seen it most of your life, Nike's slogan to just do it. Well, I'm here to tell you it's stupid! DON'T just do it. Do it right!

The great philosopher, Dirty Harry, once said, "A man's gotta know his limitations." That is a HUGE truth. Give it fifty years and you'll understand.

A couple of months ago, I decided to renovate my garage. Heeding Dirty Harry's admonishment, I hired some guys to do most of the work. Trying to be frugal, I decided to do the painting myself. I've painted before. Mostly touch-up work, but Elyse's Princess Room, the pink, and I mean pink, one is, still is, the fruit of my labor.

Anyways, since I'm not a professional and since this is MY garage in MY home, I figured I'd cut some time and not remove the shelves that are staying on the walls anyway. I'd paint around them. I'd just do it. WRONG!

In the long run, it would have saved me some time if I'd done it right and taken down the shelves, painted the wall, then put the shelves back up. I can feel it as I write. I didn't do anything uncommon, other than I didn't do it right. I twisted a certain way. I don't know what "a certain way" means, I just know I pulled something in my shoulder that hurt A LOT! For a long time. Enough that I couldn't do my usual workouts for several weeks. No excuse. Just the truth. Thanks for your concern, but I'm okay now.

Here's the rub, which it took a lot of to recoup, I knew better than to just do it. I knew what I should do to avoid hurting myself, but I chose to take the easy route, which, in the end, it was not.

INSIGHT: Here's the moral of this story: If a job's gotta get done, you'd better do it right.

Piece of pie…

I'd just assumed the duties of Deputy Director of the Family Support Center at Ramstein Air Base in Germany when I found out that we were hosting a pie-eating contest in the very near future. The ladies in charge of the event were scuffling to find judges to select the best pie. I wondered why, when we were in dire need, I hadn't been invited to be a judge, after all, I'd just become their boss. It seemed the courteous thing to invite the new boss. Yeah, I'm an Old School Kinda Guy….

I didn't say anything until we got very close to the event and still didn't have a full panel of judges. I "volunteered" to serve, at which time the ladies hosting the event gave me the "why didn't we think of that" reply and invited me to serve as a judge. I, of course, accepted.

On contest day, I was ready! I made sure not to eat too much prior to the event. I'd never done this, but I figured a full stomach wouldn't serve me well. As we started the contest, I noticed the other judges, who had done this before, taking a sliver of each pie. I wondered why so small. How would you be able to judge the taste with such a small sample? I took a normal slice of each. I knew what I was doing. NOT! After the third or fourth pie, I was full. Instead of TASTING each pie, I was eating a slice of each and filling myself not only with a lot of carbs, but a lot of sugar. I was miserable by the end of the contest. I drove through and did my part, but as soon as I figured out what I should do, I, too, started taking just a sliver of subsequent pies.

INSIGHT: The lesson here is, be careful what you pray for. I was a bit miffed that I hadn't been invited to participate in the event, but I was also not "qualified" to do it correctly. If you're going to sign up for something, make sure you know what you're signing up for and that you know how to do it right. Otherwise, you're liable to get pie in your face!

Second chances…

During my junior and senior years of high school, I spent my summers in Buena Park California, living with my uncle, Tío Rulie, and aunt, Mama Joyce, where I worked at the Lincoln and Knott Carwash.

I was having a normal day when one of the guys that worked in the detailing shop drove an old Thunderbird up to me to put on the line. He told me not to turn the engine off since it would die if I did that. They'd just steam cleaned it. It was idling pretty fast. We usually turned the engines off and left them in neutral. The line was a chain-operated contraption that pulled the cars down the line of brushes and sprayers and such. We'd have five or six cars on the line at a time.

I drove that old Thunderbird around to the line and drove it on to the next spot. I left it running as I was told. As I left the car, I slammed the door shut, which jolted the gear shift, dropping the transmission from neutral to drive. Since it went into drive and had been idling pretty fast, it jumped the tracks and kept going down the line rear-ending the car ahead, which, in turn, rear-ended the car ahead, which, did the same for six or so cars. Chaos ensued! BANG! BANG! BANG! Every car was jumping the line and running into the one in front of it! Employees, all of a sudden, appeared, I suppose to help, but more to see what the ruckus was about! After about ten minutes, all the cars had been parked out in the lot so that the managers could assess the damage. Most of the car owners were upset. Some were amused. Some had been looking to get a new car anyway so now they had an excuse for them to do so.

I walked over to the locker room, changed into my street clothes, and started to head home. As I left the locker room, the owner of the car wash, Mr Bob Burglund, came out screaming at me! “What are you doing?” he asked. He knew I was the person who had caused the problem. “I’m going home, Sir,” I replied. “You don’t have to tell me I’m fired.” “FIRED?!” he continued. “I just invested twenty thousand dollars on teaching you a lesson. Now, get your butt back to work!” he admonished me. Two weeks later, I was made an assistant manager of the place.

INSIGHT: As my boss told me when he promoted me, Mr Burglund saw potential in me. He saw me as a Shining Star. He knew that I would never make that mistake again (which I never did). He gave me a second chance to redeem myself and I did. I was totally loyal to him and to what I was being paid to do. Years later, I had plenty of opportunities to give others a second chance and I did. I was never disappointed!

Showing respect...

I was the Senior Enlisted Advisor to the Commander of the 92nd Air Refueling Wing from July 1994 to July 1996. One afternoon, I was at my desk reading through the mounds of paperwork that just seemed to appear daily. We weren't very tech-savvy then, so almost everything was done on paper. I'm nearsighted. I can read close-up without glasses, but I can't see further than about three feet without them.

Anyway, I was reading a document when I noticed a figure appear at my door. My office was rectangular in shape with my desk at the far end, about ten feet from the door. I couldn't make out who was at my doorway, but I wanted to acknowledge their presence, so I looked up for a second and said, "I'll be right with you." I went back to reading what I was reading as the person at the doorway waited patiently (I thought it was patiently.).

As I set down the document, I donned my glasses onto my face. OH, CRAP! It was the

Commander, My Boss, a Brigadier General! I'd just told him to wait on me! I immediately did what Warriors do when an officer enters the room. I stood up! I apologized! Luckily, General Voellger was one of the kindest gentlemen I've ever known. He accepted my apology, told me what he'd come to tell me (Not that he was about to promote me!), and we went about our business. Or, as Dr Stephen Covey used to say, "busy-ness."

INSIGHT: That day, I vowed to ALWAYS stand whenever ANYONE enters my office or stands at the door waiting to enter. I've practiced that to the point that it's now a habit. Let me tell you, it takes me longer, and it's a bit noisy, whenever I get up these days, but I stand up! It's the right thing to do. I'm deliberate about it. It leads to integrity. I do it every time. Regardless of rank or stature. Everyone deserves my respect and that's one way I show it.

Don't interrupt...

I'm sitting at my desk, typing on my computer, facing away from the doorway, when I, somehow, create the most powerful statement I'd ever read or conceived. I'm in the middle of writing an award submission for one of my workers. This is the killer statement! When the selection board reads this statement, they'll set all other submissions aside. I'm just beaming!

And then it happens! Carla walks into my office without knocking, comes up to my desk, and starts telling me what's on her mind! I immediately turn to hear her. It's a normal reflex.

I'm convinced that humans cannot multi-task. I'm about to prove it!

As I turn to hear what Carla is saying, I stop writing, which means I stop thinking that thought that I was in the middle of typing. I lose it. All gone. Forever....

Try as I might, I retrace my steps in order to regain the thought. Nothing. It's gone....

Afterward, I "chatted" with Carla, which means I admonished her for not waiting for me to invite her into my office. I was nice, but firm. I explained to her that I was not prepared to listen to her. In fact, my mind was on something totally different. I wasn't able to think two different thoughts at once.

INSIGHT: Interrupting is one of the worst communication habits we can develop. We expect others to listen to us, but as the messenger, we have to do our part. If we want others to listen to us, we have to give them time to get ready to do so. Don't interrupt.

Sometimes, words won't help...

My Brother, Rick, suffered with stomach and liver cancer for several years. I talked with him regularly but didn't see him as often as I should have. I always expected him to recover. He was a tough guy. I loved him deeply...still do....

I'd been back from visiting him, in Fort Worth Texas, three days, when I was in a character and leadership seminar for junior cadets at the Air Force Academy, where I taught.

About two hours into the seminar, I received a text from my baby sister, Elva, stating, "Rick's gone." He was only fifty years of age. This was in January. His fiftieth birthday was in December. I was devastated. I, again, expected Rick to recover. As I said, he was a tough guy, after all.

I was seated with a half dozen football players. The team's linemen. Big bulky guys. Average height of six foot four. Average weight of two hundred fifty pounds. One of the largest was

Nick Charles. He played guard. He was sitting next to me.

Nick saw me look at my iPhone as I read the message. As I looked up, he looked at me and asked, “Are you okay, Chief?” “No, Nick,” I replied as I showed him my phone. “Come here, Chief,” he told me as he had me stand next to him.

And he hugged me. No words. Didn’t need any. Wouldn’t have done any good. Wouldn’t have lightened my pain. He just hugged me. I can still feel the sense of being surrounded by his love and his protection.

INSIGHT: Most often, when we experience a friend having lost a loved one, we immediately search for the right words to say. Don’t! Most often, there are no words that will comfort that friend as much as a sincere hug. Just hug….

Sometimes, ya gotta lie...

I was on a KC-135 aircraft with a group of USAF Academy Cadets, the Drum & Bugle Corps to be exact. We were returning from a football game. As we made our landing approach, I noticed that we, instead of landing, circled around. A KC-135 has only a couple of very small windows, so no one really knew we'd done that. I'd served in a KC-135 wing and taken plenty of flights, so I knew the feeling of starting to descend then rising again, then circling. As we pulled up to circle, I noticed the crew scrambling a bit. They were rushing toward the landing gear. I hadn't heard the landing gear drop but I thought it was that I'd been talking and just hadn't been paying attention. As we began to descend again, the crew began scrambling even more. Evidently, the landing gear was stuck. I began to worry. I could envision us having to land without landing gear. I could see the foam the airport folks would have to lay down for us to land on our belly. We'd have to quickly scuttle from the aircraft if that were to happen. I could see how we'd have to guide the Cadets. I was making a plan as to how I'd do what I needed to do to save and protect my charges.

As we made our second circle, some of the Cadets noticed that something was not right. A few of them, at different times, came up to me and asked me if everything was okay. Sorry, but I had to lie. I had to tell them everything was okay. I truly believed that we'd land safely, but not as usual. What was I to do? Tell them that we might die? Not on my watch! There's hope. Our Airmen can do unbelievable things! They'd save us.

Well, after seeing the crew practice innovation and resourcefulness, they got the gear unstuck and we landed safely, after circling only three times. After all the lessons I'd taught my Cadets about having integrity, I had to lie to them to protect them from freaking out. I reminded myself that integrity is about doing the right thing and this time it was lying. I hope I never have to do that again. It didn't feel right, but it was right.

INSIGHT: Sometimes, keeping the truth from someone is the best thing to do.

Do it, or I'll kill you!

Back in the day, we were often "trained" through fear. I remember my first assignment with the 15th Air Force Band at March Air Force Base in California. Within days of arriving, the three of us Young Airmen, at that time, Mike Mormon, Mike Paulson, and I, were taken outside by Staff Sergeant Rick Rose, the Duty Sergeant, for him to tell us what our additional duties were.

"Airman Mormon, your job is to…." I don't recall, but it was much easier than mine, I assure you. "Airman Paulson, YOUR job is to…." Whatever. Then, "Airman Vásquez, your job is to keep this sidewalk clean." There was a huge eucalyptus tree right next to the Band building. The building and the tree were separated by a sidewalk, which was now my responsibility to keep clean. That tree shed its bark like it was going out of style. That sidewalk was constantly strewn with eucalyptus bark. I noticed it the first time I walked up to the building.

Sergeant Rose gave me specific instructions. “Here, Vásquez, take this broom and sweep that sidewalk. I don’t want to see any bark on it. Ever.” “WHAT?” I thought. Evidently, I wasn’t just thinking it, I was saying it, and more. “WHAT? You want me to sweep that sidewalk every day and keep it clean every day, all day long? Are you crazy (I thought I was just thinking.)?” In those days, it was a big thing to have gone to college prior to enlisting. You graduated from Basic Training as a two-striper, an Airman First Class, vice an Airman Basic no-striper. I’d gone to New Mexico State University for three semesters, which made me eligible for the extra stripes.

“Look, Sergeant Rose,” I told him, evidently, “I’m an Airman First Class. I went to college. I’m a musician. I’m a bassoonist. I don’t do that!” Again, I was thinking I was thinking, not saying. Man, did that set Sergeant Rose off! I can still feel it...he sticks his finger in my chest and tells me, “Do it or I’ll kill you!” Let me just say that that sidewalk was the cleanest sidewalk on base for as long as I was there.

INSIGHT: The lesson here is that, back in the day, that’s how we were trained. That doesn’t work anymore. You can't even think those thoughts. But that's okay. As we evolve, we have to devise different, hopefully better, ways of leading people. Keep evolving….

Tell the kids I love them…

I was sitting in the principal's office waiting to have lunch with my nine-year-old granddaughter, Button, when, as I looked around the artifacts the principal's assistant had displayed all over the walls of the waiting area, I noticed a sign that read "Tell the kids I love them. God." WOW! I'm a Chief. Retired, but still a Chief. Chiefs don't cry, but sometimes our eyeballs sweat. They perspired a little as I read that sign. It touched me deeply.

I strive, and profess, to be a loving man. I understand love as unconditional giving. Deb says I, sometimes, go too far with practicing love, especially with strangers. As I see it, I'll give anyone in need of what I have whatever I have, as long as I have it and can give it. I'm blessed with so much!

Giving love doesn't cost a cent, most times. Especially the kind of love that, I believe, God would have us give. I love kids. All of them! When I'm given the opportunity to be near or among them, I often ponder what they might be thinking or seeing or needing. I'm grateful for every single kid I've ever seen. Each one was endowed by the Creator with certain talents and gifts that our world needs.

I'm blessed with six grandchildren. They range in age from 20 to 4. Each is different and beautiful. They truly are. One thing they all would agree to, though, is that telling them, not as God would, but as we humans would, isn't enough. We have to show them. We have to be respectful of them individually and appreciate them personally. That, I think, is how we deliver God's message to them.

INSIGHT: The lesson? Tell the kids I love them. God

Thank you for your service, Sam…

I'm recalling the week that we celebrated the Air Force's 73rd birthday. That week, a very close friend of mine posted a picture of the Air Force Memorial in Washington DC, which reminded me of what I'm about to share with you. Thank you, mí hermanita, Wendy Thi!

Years ago, when I was teaching youngsters how to become Leaders of Character, I used to show them a video clip of an Army Private First Class who was killed by an IED in Baghdad by the name of Sam Huff. She was only 18 years old when she died. It still hurts me deeply to think of her making the ultimate sacrifice. She was just beginning her life.

I was in DC just before the Air Force Memorial was opened, so it must have been the Fall of 2006. Sam was on my mind, so I went to Arlington Cemetery to visit her grave. I can still feel the overwhelming sadness and pride that overcame me when I found her marker (I could take you there right now.). I told her about the impact her story was having on hundreds of Airmen who would soon be leading the United States Air Force. As I laid my coin on her stone I was drawn to look up and to my left, her right, to see the three spires of the Air Force Memorial reaching toward the blue sky. It was as though she'd guided my sight and my spirit toward it as we both acknowledged that we Warriors are all in this together, even if we serve in different services.

INSIGHT: Chiefs don't cry, but sometimes our eyeballs sweat. They did. A LOT! But it was all so good with my soul as I said so long to a courageous young lady from Tucson Arizona who was willing to die so that we, you included, could live a great life in the United States of America! Thank you for your service, Sam.... Here's a toast....

How do you know?

I served (indirectly) under General Robert Rutherford while he was the Commander of Air Mobility Command. He was a no-nonsense kinda guy. Although I'm told by folks who worked for him directly that he wasn't as physically large as I remember, I don't know anyone who ever saw him smile. Not that he was mean or anything of the sort. He was just a very serious man.

He used to go around his bases, checking on his people and his wings' operations. He'd get to Base X and as he was being shown around would ask people, "How are things going?" Imagine that. A four-star commander asking how things are going. How do you answer that? "Great, General!" Regardless of the truth, you'd answer in that way.

His follow-up question would hit people between the eyes. "How do you know?" Duh…crickets…bumfoozled looks….

I've grown to value that question, "How do you know?" Call it measurement, assessment, whatever term you want to use, but if you don't have some way of analyzing how you're doing, you won't get any better, assuming you want to improve.

INSIGHT: Stephen Covey taught me that "as long as you keep doing what you're doing, you'll keep getting what you're getting." If that's good for you, more power to you. But if you intend to get better, it pays to know where you're at, to begin with.

The cut-off…

Our attitude can have a huge effect on others. It's contagious. Sometimes, we don't even realize the effect it can have.

I'm stationed in Germany. I'm on my way to work, which is about twenty-six miles from my home. I'm driving fairly fast. I'm driving my BMW. About 100 MPH, when a German, driving an Audi, passes me. Just as he passes me, he decides to get into my lane, and slows down, cutting me off. I'm not a happy camper.

As I get to the shop, my Administrative Assistant, Barb, welcomes me by asking, "How are you doing, Chief?" I'm still reeling from that Audi driver cutting me off, so I express myself rather animatedly, surely, emotionally. Barb's a great person so she listens to me empathically and almost consoles me. I go upstairs to get some work done. The day goes on….

At the end of the day, I'm on my way out to my car when I see several Airmen along the way.

The first one I see tells me he heard about my morning encounter and sympathizes with me. The next one does the same, giving me advice on how to handle it next time. You have to understand that there are 40,000 people assigned to this base and 39,956 of them know me. As I make my way to my car, every single person I see has, evidently, heard about what happened to me this morning, and everyone sympathizes with me. In fact, they seem to be as ticked off as I was when it happened. Interestingly, I'd forgotten about it by now.

INSIGHT: As I said above, our attitude is contagious, and we often don't even realize its effect on others. I have no idea how many people heard about my morning encounter, but I do know that the attitude I developed because of it, affected a lot of people. We can choose our attitude. Choose wisely.

The language of caring...

On my way back to Spokane Washington from St Louis Missouri, I got diverted to San Francisco due to severe weather between Spokane and St Louis. The flight to San Francisco International Airport was nice and smooth. Although the airlines wasn't much help. I arrived too late to take the last flight to Spokane, so the airline booked me for a flight the next day. They put me up for the night, in a hotel of their choice. I had to look around to find someone to tell me where to go to get to whatever hotel I'd be spending the night at. I finally did find an airline representative to guide me. She gave me a voucher and told me to wait for the shuttle at a particular door. I did as I was told. I'm an Airman, after all. I follow orders.

About twenty minutes later, I see the same airline rep, along with a lady holding a baby, coming toward me. She approaches me and asks me, "Can you help this lady?" You know how you call a store, and they ask you if you can hold and put you on hold before you can say no? That's, more or less, what happens. She asks me that question as she gives me a card and takes off somewhere else. Very quickly!

The card reads something like, “Hello. I’m Vietnamese. I don’t speak English. I’ll be staying at the same hotel you’re going to. Please help me check in.” Okay. So, I assume that the airline rep picked me to help her because of my charming personality, not the fact that I was the only other passenger waiting to go to the hotel.

I did a couple of tours in the Far East, as we called it then. I flew in and out of Vietnam a couple of times, as I was told. But I don’t speak a word of the language. I do, though, know a Banco Filipino radio ad verbatim, if you ever want to hear it.

Since I didn’t speak her language and she didn’t speak mine, I decided to speak to her in Spanish. Hey, she didn’t know the difference! Amazingly, I got us all checked in after we found the right shuttle to take us to the hotel. I’m not sure my verbal attempts to communicate with her worked as much as my gestures. After we checked in and I guided her to the elevator and gestured which floor she should get off on and her room number, she bowed to me, and I returned her bow. I think we were both appreciating each other.

The next morning, I asked the hotel clerk if the lady and the baby were still there, but they'd already left. To this day, I wonder if I just made all of this up in my imagination. Actually, I kept the card for several years. Some of it did, for sure, happen.

INSIGHT: The lesson here is that if you get stuck at an airport, stay away from airline reps. Okay, that's not it. The lesson is, be helpful regardless of your language. I can't imagine being in a place where I don't know at least a little bit of the local language. I even got through six and a half years in Germany, not knowing the language. I always knew when I was being yelled at, though.

Being human is not governed by how we speak. It's always governed by how we act, which is how we communicate. What we do, speaks louder than what we say.

I love you, Dad...

I remember this event so vividly.... Dad wasn't the most touchy-feely guy in the world. He grew up tough. And the culture that permeated the years he raised me were not what they're like now. I don't remember him ever saying, "I love you" until this event.

I was stationed in the Philippines. About 1974 is when this happened. Calling home was a very expensive gift I gave myself and my parents as often as I could. Which, as a young Airman, wasn't often enough.

As I recall, the cost was $25 per minute to call from Clark Air Base to Deming New Mexico. Yeah, 25 DOLLARS per minute! I wrote a lot of letters then but saved my money in order to make a call home once in a while. I believe it was around Thanksgiving when Dad and I had this conversation.

I'd saved up $100 to make this call. I'd written Mom and Dad to coordinate the call and to tell them we only had four minutes to chat, to make sure they knew what they wanted to say. You have any idea how long four minutes is? Too long when you don't want it and too short when you do.

I talked with Mom for a couple of minutes (literally) and then it was Dad's turn to talk. We chatted for a minute and almost two when something compelled me to say, "I love you, Dad." Silence on the other end of the phone. Dad! Do you know how much this is costing me?! Of course, he didn't. The seconds felt like hours. He finally replied, "I love you, Bob." From that day on every single one of our conversations ended with me saying, "I love you, Dad." And him replying, "I love you, Bob." How I wish we could have those conversations again. I lost my dad in 2015.

INSIGHT: Always tell those you love that you do as often as you can. Better yet, show them.

I love you, Dad!

What humility looks like…

I wasn't a General Tommy Franks fan until this happened.

I'm at the College of the Ozarks, where my daughter attends school. I'm one of the speakers at their annual Character Symposium. Our host gathers all of the speakers and other distinguished guests to take the traditional group photo. We're all herded outside and told where to stand. I'm posted at the back corner of the group, due to my height. Or so they tell me. As I'm standing there, waiting, General Tommy Franks, a couple of his assistants in tow, joins me. As I turn toward him and notice it's him, I shake his hand as I introduce myself and drop some papers I had in my other hand. As I start to bend down to retrieve my papers, the general beats me to it. He captures all of my papers and returns them to me saying, "Here you go, Chief."

Now, that may not seem like a big thing to do, but here's a four-star general, thirty-eight-year veteran, former Commander of the United States Central Command (That's a big deal to us military folk!), picking up the papers that I, so clumsily, dropped, when he could have had his assistants do so. He, personally, took it upon himself to serve me! That is what I consider humility, which he exuded the entire time we were together at that event. Chances are that's the nature of his character.

INSIGHT: Humility is a critical trait of a leader. We don't talk about that much and we, surely, don't teach it. Maybe because it's near impossible to teach. The only way to teach it is to show it. The only way to hone it is to practice it. I'm a General Tommy Franks fan now. I acknowledge that with all humility.

The motivation myth…

People constantly ask me how to motivate people. They're usually not satisfied with my answer that you can't. Here's an example.

My family and I are at the mall (a place with a lot of stores where people used to go to shop for stuff) in Riverside, California. It used to be one of our favorite pastimes. Now, we just order from amazon. Anyway, Deb, Tesa, Elyse, and I are at the mall, just looking around. We're in some store in which Elyse, about six-years-old, decides she wants something. She HAS to have it! "No you don't," I tell her. She has a fit! She's crying and screaming at the top of her voice that she wants it! I can't do anything but take her out of the store. It was pretty embarrassing.

I take Elyse out to the mall and start to motivate her to stop her crying. She continues on. At the time, I was the Commandant of the Noncommissioned Officer Academy at March Air Force Base, right outside of Riverside. I was THE expert on leadership, as per my title, which, of course, included motivation. I give her my best stuff, citing motivation development theories and sharing examples, etc. She has nothing to do with it.

Eventually, I get so frustrated that I decide to take her out to the car. As I take her arm, she goes limp. COMPLETELY limp! I catch her before she falls to the ground. Have you ever noticed that many drunk drivers who collide with something, including another vehicle, often survive without a scratch? Why is that? Because they're so limp, they go with the flow. What will hurt you worse than anything when you take a fall is that your first inclination is to fight it, to become rigid. That's how you break a wrist, an arm, or some other limb. As I try to pick her up, Elyse turns into a glob of Jello. She's not motivated to do what I wanted her to do. She's already motivated to do what SHE wants to do and does!

INSIGHT: We CAN'T motivate people! They're already motivated! To do what they want to do or are already doing. The best we can do is INSPIRE them to motivate themselves to do what we want them to do, and, hopefully, what they SHOULD do. If you're looking for advice on how to control your kid when what I shared with you happens, I have none. Sorry. What I did with Elyse was seek to understand what motivated her and offered her a way out. It worked. We both loved that BMW!

What right looks like…

I served at the United States Air Force Academy from 2002 to 2022. On this particular day, the Freshmen, we call them Four-degrees, are going through their Rite of Passage exercise we call Recognition.

I'm out on the Terrazzo observing the Four-degrees excel at what they do best…becoming Leaders of Character. I'm over by the Hill where they're formed up doing flutter kicks. As I'm standing there, an Upperclassman walks up to me and we start to chat. As we're talking, we both notice a Four-degree who is having a really hard time doing the kicks. So hard a time that she quits. Just stops. You can see she's given all she's got, but it's not enough.

As the Upperclassman sees her, he says, "Excuse me, Chief," and walks toward the Four-degree. "Oh NO!" I'm thinking. "Bad things are about to happen to My Four-degree!"

The Upperclassman lies down next to the Four-degree and starts to do flutter kicks along with all the other Four-degrees. The Four-degree sees him, takes a deep breath, then starts to do the kicks again. They both continue until the trainer-in-charge blows his whistle to move on to the next station.

The Four-degree can barely get up. She's struggling. The Upperclassman grabs her by the arm and they both head off to the next station. Curious, I follow them to see how long the Upperclassman assists the Four-degree. He helps her for at least two more stations. I, eventually, can't keep up with them. They continue driving on.

INSIGHT: As Dr Ken Blanchard says, "none of us is as good as all of us." Although all success starts with the individual, how much that individual is willing to commit to that success, there's a point, usually a short one, where we have to realize that we need others to help us succeed. That's called humility. When we acknowledge the importance of others, we can make ourselves even better.

The punch bowl…

I was stationed at Keesler Air Force Base in Mississippi for a few years. I was very involved with the Hispanic Heritage Club on base. One of its members and a gentleman assigned to the base was Dr José Vásquez. He was a good man. We didn't have the internet or email yet, so I used to get his mail constantly. I used to take it to him regularly. Unless we were face-to-face, we were constantly mistaken for each other.

One evening, Deb and I are chilling at home when the phone on the wall that had a long curly cord rang. As I answer it, a female voice on the other end starts the conversation with, "Is this José Vásquez?" I'm known as "Bob" by most people. But the folks in the Hispanic Heritage Club all know me as "José." I go by Bob because it's a nickname for my middle name, Roberto, and, mostly, because people offend me by not pronouncing "José" correctly. You might be surprised how some people can mispronounce it.

Anyway, back to the phone call. “Is this José Vásquez?”

“Yes, it is,” I reply.

“Good!” the obviously-Hispanic voice on the other end confirms. “When you come over to the dinner tonight, can you bring your punchbowl?” she asks. Punchbowl? I didn’t know we had one. I put my hand over the voice receiver and ask Deb, “Do we have a punchbowl?” She shakes her head. “Someone was supposed to bring theirs,” the voice continues, “but they just called and said they can’t make it. So please bring yours.”

“Uuuuh, we don’t have a punchbowl. I’m sorry,” I tell the voice on the other end of this conversation.

“But you just told me yesterday that you have one. This is José Vásquez, isn’t it?!”

“Yes, it is. Are you looking for Dr Vásquez or Master Sergeant Vásquez?” I ask.

“Dr Vásquez,” the now-agitated voice replies.

"This is Master Sergeant Vásquez," I correct her.

"Oh, okay. So, do you have a punchbowl?!"

INSIGHT: The lesson in this story is that we often start and continue conversations not knowing who we're talking with. Technology has done that to us. Instead of helping us communicate better, it, sometimes, enables us to miscommunicate, efficiently.

BTW, we finally did get a punchbowl.

The open-door policy…

I'd just sewn on Chief and had the honor of going back to where I started, March Air Force Base, near Riverside, California. I hadn't been there long when my commander, Major Denny Layendecker, summons me to his office.

"Chief," he says, "please, take my office door down."

"I'm sorry, Sir, (I didn't think I'd heard him correctly), did you say take your office door down?"

"Yes, Chief. Take my door down," he replies.

"Surely, Sir, but may I ask why?" I ask, bumfoozled.

"I want to have an open-door policy. I want the troops to know they can come in and talk with me whenever they want."

"Okay, Sir," I acquiesce.

Now, as I said, I'd just sewn on my Chief stripes. Chiefs don't do that type of work. They have Duty Sergeants who are training to be Chiefs one day. So, I find the Duty Sergeant and ask him to take the Major's door down. Yeah, I had to explain why. It was done.

About a month later I'm chatting with the boss and the topic of listening comes up. I ask him, "So, Sir, how's it going with your door down? Are people coming by to chat? You made that announcement in your Commander's Call."

"Chief," he looks at me, and in a bit of a saddened tone replies, "Not one person has come by."

"Hmmmm...," I murmur. "Your intent was to hear their thoughts, their problems, and learn about them, right, Sir?" He agrees. "Here's a suggestion. How about YOU go out there and hang out with THEM. Where they live and work. (The shop was on the first floor of a dorm. The Airmen lived on the floors above.)"

"I hadn't thought of that, Chief," he perks up. "I'll try it!"

A few days later he comes to my office and excitedly asks me, “Chief, did you know that Carmen is pregnant, and that Jerry’s mom is coming to visit him next week, and that Smitty inherited a lot of money?” he goes on and on.

“Yes, Sir, I know all that.” I reply.

“How did you know all that?” he asks, almost surprised.

“Because I do what you’ve been doing, Sir. I hang out with the troops trying to figure out what makes them tick so that when I make recommendations to you they’re based on what’s important to accomplishing the mission and to taking care of our people. We can’t be excellent until we know what that means and how we can do it.”

“Thanks, Chief! By the way, can you put my door back up?” he asks.

“Can do easy, Sir,” I answer. And then I went and found the Duty Sergeant.

INSIGHT: Communication is key to success and even to having a good life. It requires listening, though, not just telling. Effective listening is done by listening with our hearts and eyes as well as with our ears. But we have to be there. We have to see and feel what's happening to truly understand. Listening is an active behavior. A leader can't wait or expect for her followers to come to her to talk, she has to find them, too!

They're actually watching...

I recently congratulated one of my Facebook Friends, and a protégé of longer than I even know, on his educational accomplishment of completing his degree. In a science. To me that's huge. I don't do science. I don't believe in it. I do art. That's true reality. Anyways, his reply to my congratulatory note was, "Thanks Chief!!! I learned it from watching you." Oh, CRAP! He saw ME doing what he accomplished?! Luckily, he succeeded. And I suppose I did too, in a way.

My Princess Number One, Nieves, who is now about to turn twenty years of age, once, as a five-year-old, impressed me with doing something out of the ordinary, in a good way. I asked her why she did that thing she did the way she did it. She looked at me with an almost surprising look as she replied, "Opa, because you do!" Another "Oh, CRAP!" moment!

I was having dinner the other night when I did something that reminded me of someone and for a few minutes I couldn't recall where that came from until I realized that I had just done what my dad used to do at the dinner table. In fact, I could see his hands doing it, only from MY perspective. Freaked me out!

My point in all of this, as I relayed it to another protégé this week, is that we learn by watching. We teach by showing. We model what our protégés will eventually do. The "Oh, CRAP!" part is that we need to be aware of that and do what we do the right way and for the right reasons. And when we do it right, we can accept a little bit of the credit from others' success. It's a mutual success, in a way. And that's all good.

INSIGHT: Do right so that others will learn to do right from you. That's the cycle of life. Now, if you'll excuse me, I have to go read a physics book I've been meaning to read for a while. Who knowns, maybe you will, too!

Honesty IS the best policy…

I drive BMWs. Plural. They're the ultimate driving machine. No, really, they are! I've had at least 15 of them. Not all at once. The most I've had at once is three. Now, I'm not boasting. I've just been blessed.

I'd had my first Beemer for a couple of years when, for some unknown reason, I considered buying a Volvo. I'd seen an ad in the paper about a new model that had just been introduced. I'd never driven one before. In fact, I'd never even touched one before. So, I took Deb with me to visit the Volvo dealer. We were living in Germany at the time.

As we arrived at the Volvo dealership a very nice young man with very red hair came up to us, introduced himself, and offered to help us. We accepted his invitation, telling him about our interest, MY interest, in their new model. After asking a few questions, Red (that's what I called him) said something that blew me away. "You don't want this car," he suggested. I couldn't believe my ears. A car salesman convincing me that I didn't need HIS product?! No way! That's unheard of!

That's unAmerican! Oh, yeah. He wasn't American. After making his case for me not wanting to buy his car, I didn't. Deb and I eventually went on our merry way…in my BMW.

There were about 20 thousand people in the Kaiserslautern Military Community at the time. I figure that 19,980 of them knew me. Right around there. I told as many with whom I had "where to buy a new car" discussions about Red and him not selling me his car. Last I heard, BMW had hired him and he was doing very very well, selling a lot of BMWs.

INSIGHT: Here's my point. Red could, surely, have sold me his car. I wouldn't have been happy with it. I would have told all of my friends and acquaintances about the experience. Actually, I did. Honesty IS the best policy. Red's success was based on his honesty. As I talked cars with others and his name came up, every single person who'd known him said the same thing. You can trust him.

Whether it's in life or in sales. Be honest. It works!

Unconditional love…

My Dad taught me to love, the verb, not the noun. Love is unconditional giving. I often hear or read religious scholars refer to unconditional love (agape). Love IS unconditional! When you give from your heart without expecting anything in return, that's love!

Dad practiced that with his family and even with strangers. He practiced caring, which is a behavior of love, in all that he did. At his funeral a friend came up and shared with me that when he took his little daughter to the store where Dad was a meat cutter, the first thing Dad would do is put down what he was doing and come around to the front of his frig to talk with his daughter, asking her how she was, expressing his care for her and her health. Dad cared, and he showed it.

Unconditional giving. That's love! I remember one evening when my brother, Rick, brought a friend over for dinner. I believe his name was Joey. Joey was wearing the new faddish pants, jeans that had tears in them. Dad came home. We had dinner. Throughout dinner, Dad kept eyeing Joey's pants. He never said anything to Joey so as not to embarrass him. After dinner, Dad got up, opened his wallet, and took all of the money he had in it and gave it to Mom. "Get him a good pair of jeans," he told her. I'm not sure how much money Dad had given Mom, but I know that it wasn't even close to the amount Joey had paid for those jeans. It was all he had, and he expected nothing in return. It was an unconditional gift.

INSIGHT: Love is a force, an energy that, once produced, cannot be destroyed. Dad gave all he had, every day, to whoever needed it. He expressed that force in everything he did. I lost my Dad a few years ago, but I'll never lose his love. It's unconditional.

Weathering the storm…

Two important parts to this story. I drive BMWs. I live in Colorado. I started driving BMWs in Germany. Man, it was fun speeding down the autobahn at 130 MPH in the ultimate driving machine. Had to move over to the right lane, though, whenever a German driver came up behind me and flashed her headlights. I'm good with that. I'm a good guest. At last count, I've owned 15 BMWs. I kinda dig them…. The weather in Colorado can, and often does, change every five minutes or so. From one extreme to the other. You learn to live with it or put up with it….

So, I'm at work, about ready to call it a day. The weather has been normal…hot, cold, wet, dry, windy, calm…normal for Colorado. As I gather my stuff to head home, I call Deb to tell her I'll soon be on my way. She admonishes me to check the weather. There's a storm brewing that might have hail within it. Hail and BMWs don't mix well.

As I walk out of the building, it's sprinkling. I look up at the sky. The dark clouds coming my way are still on the other side of the mountains. I should be good to get home before the storm hits. Maybe I can drive faster than it. I'm about a mile into my seven-mile trek home when it, all of a sudden, starts to rain huge drops. Deb calls me on my iPhone and warns me not to come home yet. The storm is almost upon where I work. Too late! Within minutes of our conversation, the hail starts to descend upon me and everyone else trying to beat the weather home. I'm in the middle of nowhere when golf ball-size hail starts to bombard my X3. I, and everyone else, am looking for cover. There's none in sight. It's very difficult to tell if the hail is hurting my car, but it's loud and I can feel every bump. I have no other way to go but forward. Now, I can't go faster, that would be really dangerous. I just have to weather the storm. And hope that it moves quickly.

As I left the base, I found a bridge under which I, and a bunch of other drivers, found refuge until the storm passed, which it did as quickly as it came upon us. My car was hurt. Almost as much as my ego. About six thousand dollars' worth. But I got home safely. From then on, I made sure to check every weather channel I could find, and I called Deb to give me the all-clear BEFORE I left the job. It hurt, but it was a good lesson to learn.

INSIGHT: Sometimes, you have to keep moving forward even when it hurts. Not doing so may hurt more. The best we can do is take the hits and continue cautiously and safely. There's no other way. Keep driving!

American Warriors...

I'm writing this on Veteran's Day 2020. In a world that's gone crazy, one thing that remains consistent is the American Warrior and the American Warrior Family. I entered the United States Air Force in 1972, not because I was a patriot, in fact, for the opposite reason. I was trying to avoid the draft. I didn't want to go to Viet Nam as a Soldier. I was blessed to have been selected to serve in the Air Force's Band program in which I served for 24 of the almost-31 years I ended up serving. I NEVER intended to serve more than four years, but wisdom serendipitously hit me between the eyes. I served my country for fifty years.

What changed my mind about staying that long is, as I said, the American Warrior and the American Warrior Family. With all due respect to all Americans, American Warriors are a special breed. The purpose we Warriors serve is to protect our country and our fellow Americans. We don't care much about the color of their skin, their cultural differences, or their religion. If they're Americans, we protect them. We even protect our Allies. That makes us a special breed. I say breed because as much as we call what we do a profession, it's really an identity. I often ask my protégés and students when our duty expires. It doesn't. NEVER! Once a Warrior, always a Warrior. Even after we stop wearing our uniform daily.

And let me not forget the sacrifice our families make. They don't get paid a dime for all they put up with. They persevere through countless moves. They sacrifice great jobs to relocate every three or four years. Many, unfortunately, have to live the rest of their lives without the loved one who made the ultimate sacrifice to protect and defend their country.

INSIGHT: I'm proud to be a Veteran of the Profession of Arms! I'm just as proud of my family who has been right beside me the entire way. Each of us bleeds red, white, and blue. Thank you, my brothers and sisters, for your unselfish service.... HEIRPOWER!

Welcome home!

My daughter, Elyse, graduated from the College of the Ozarks in Branson, Missouri. One summer, as we delivered her to the campus, we found out that they'd just opened the Dick Clark's American Bandstand Theater in town.

You're probably too young to know, but back in the day, all of us youngsters used to watch American Bandstand religiously. It was on television. In black and white. It came on Saturday afternoons. It was our connection to what was hip in the music world. It was hosted by the ageless Dick Clark. Ask your grandparents about the show and Dick Clark.

So, the Dick Clark's American Bandstand Theater opens up in Branson the same weekend we're in Branson. The headliner is Paul Revere and the Raiders, one my most favoritest groups ever! I MUST see them! I ask Deb, Elyse, and her future husband, Jeremy, to join me. I cop some good tickets and we attend the show.

During the show, right in the middle, Paul Revere has EVERYTHING going on on stage that you can imagine, singers, dancers, smoke, fireworks, horses, you name it. Horses was a little weird, but HEY! This is Paul Revere! As all of these things are happening at once, he, all-of-a-sudden, stands up from his keyboard and shouts, "STOP!" And everything does. And then he says, "All of you Vietnam War Vets, please stand up." There were half a dozen of us in the audience. We all stood up. And then Paul Revere says, "Welcome home!" He then returns to the show. Sorry. I get a little choked up when I recall this.

After the show all of us vets got together in the foyer. It's what vets do. We get together to help each other. All of us vets agreed that day that it was the first time anyone had ever welcomed us home. Some of my Veteran Brothers we in wheelchairs, some were supported by canes, all were older than I. They called me Sonny. When we returned from the Far East back decades ago, none of us had received a welcome of any sort. In most cases, it had been the opposite. We were spat on, called all sorts of obscenities, baby killers, and just treated disrespectfully. Lucky, that changed….

INSIGHT: The lesson here is that many have sacrificed themselves, their health, mental and physical, to protect this great country of ours. Consider thanking them when you get the opportunity. As one of my new friends said that day, “We did it for you!”

Born or made?

The question of whether leaders are born or made has been argued since someone came up with the word "leader." And we'll probably continue to argue that question.

Are leaders born or made? The obvious answer, of course, is YES! We were all born and we each lead, at least ourselves. But some people were born with the innate talents to lead others. Evidently, I'm one of those. I've always led. Even when I didn't want to.

My family and I had just moved to Germany to serve there while still in the United States Air Force. We were living in Lohnsfeld, near Sembach Air Base. I'd been involved with the Base Chapel at several of the bases where we'd previously been assigned. I heard that the Protestant Chapel Leadership Team was meeting the coming Tuesday, so I decided to attend and offer my help. I knew no one in the community. Not a person. As I said, we'd just arrived.

All of the folks at the meeting greeted me with respect and courtesy. They even invited me to sit at the table with them. I soon found out that most of the leadership team was returning to the States, so they were looking for new leaders. Being the new guy, I couldn't refer anyone to any of the positions they were intending to fill. I just didn't know anyone yet.

I had to go pee, so I excused myself and went to the bathroom. When I returned, the entire group was smiling. They'd just elected me the new president! I didn't even know these people and they made me their leader. Just like that!

I ended up leading that group for about six years. Every year or two I would suggest getting new leadership in to give us a fresh perspective and give others the opportunity to lead. They wouldn't hear of it. "You're our leader, Chief," they would admonish me. I had to leave the country to quit that job. Don't get me wrong. I enjoyed working with all of the folks involved in that small church. We were like family.

INSIGHT: Here's what I've learned. Leading is about making a choice to help someone. That someone is sometimes me. That's where it starts. Mostly, it's someone else. It's about stepping up to that opportunity when someone needs help or support. If you're a born leader, don't fight it, just do it as best you can. If you aspire to lead, just do it as best you can. Regardless of whether you were born to lead or are learning to do so, help each other. That's all we can do.

You don't understand the struggle...

My Princess Number One, our first granddaughter, Nieves, and I were visiting my parents in New Mexico, as was our annual tradition. We stayed with Mom and Dad. They were Old School Types. In other words, they had no internet in the house. Nieves was about twelve at the time, almost thirteen. She was "almost thirteen" since she was about five years old. Not having internet was driving her nuts! She couldn't communicate with her friends back home. She had a phone, but no internet! How unAmerican! I was almost okay without it, but not quite. I decided to help calm Nieves down by taking her to breakfast at McDonald's where they DID have internet. As we walked into the restaurant, I was giving Nieves a hard time about her need for online connection. She shut me down quickly when she said, "Opa, you don't understand the struggle." I had no reply to her statement. She was right! I didn't.

I'm not sure when it happens, but it does. We

all evolve to a point where we either think or, usually, say, "Been there. Done that." But that's not so. We've NEVER been here, nor done this. Life evolves constantly and, nowadays, faster than ever.

We NEVER understand the struggle. Not their struggle. We barely understand OUR struggle. We don't know what someone else has been through, even if we've raised them. How many times have you wondered where "that" came from when your child surprises you with a "crazy" thought or behavior? By the way, it probably came from you.

INSIGHT: We don't understand their struggle, whoever you're trying to understand, or worse, control. The best we can do is practice empathy. Try to understand that they feel as they do. Don't judge. You're not worthy of that. Thanks for the lesson, Nieves!

What hasn't changed...

I joined the Air Force in 1972. Life was different then. Technology was different then. What is the same, though, is the importance of connecting with others.

The first person I met, on the bus from the San Antonio Airport to Lackland Air Force Base Texas, where we both "enjoyed" Basic Training, was one Andy Ducharm. He was a red-headed, loud-but-funny, kid from New Hamsha. Otherwise known as New Hampshire. He talked funny. I'd never heard anyone talk like Andy. Hey! I'm from New Mexico. No one speaks with an accent there!

One event we Jeeps were always anxious for was smoke breaks. "Smoke 'em if ya got 'em," was a welcome order from our Military Training Instructor, Technical Sergeant Partridge, who smoked like a chimney. If we were in the dorm, we'd be allowed to head to the break area to smoke. Many of us didn't smoke, but we faked it, just so that we could get a break. Those who didn't, stayed in the dorm scrubbing the bathroom tile with their toothbrushes.

The break area was equipped with what is now unseen technology, pay phones. There were about a dozen. There were way more of us than them. Everyone who could afford to make collect calls, where the other party paid for the call (We weren't allowed to have cash and calling cards hadn't been invented yet.) would line up for their turn to call home. Andy ALWAYS called home. As he ended his conversation with his parents, he'd make an announcement (I can see him now as if it just happened.), "Anyone wanna tawlk to New Hamsha?!" Many of us would line up to talk with whomever was on the other end of the phone call, usually Andy's mom. We'd all say hello and wish her blessings as she blessed us.

INSIGHT: Connections and connecting is what makes us human. I can still feel how good it felt to touch the outside world that we were being trained to protect. I lost track of Andy after Basic. I've tried many times to find him, with no success. Andy, if you're out there and read this, please call me...collect!

What is hip?

I recently posted the eternal question on my FaceBook page, that is, "What is hip?" You have to be a Tower of Power fan to really understand the question and to have the right answer. Tower of Power, if you're too young to know, is, since 1968, the baddest funk band in the world. Google the group or go to iTunes to listen to them. I guarantee you you'll be on your feet fast!

Anyways, one of my FB Friends, Cheryl, wrote back with the answer, "Hip is being comfortable in your own skin." I like that, but the answer, as the tune goes, is "Sometimes hipness is what it aint." There's some deep philosophy there. In both statements.

Of course, I'll agree with my friend, Cheryl, that being comfortable in your own skin SHOULD BE hip, but is that so? Especially when you're young?

I remember having a conversation with my 10-almost-13-year-old granddaughter, Nieves, about her vision of one day being Hannah Montana. Try as I might, I never got through. She was going to be Hannah Montana, in other words, in someone else's skin. My argument was for her to be Nieves Vásquez! She could be a better her than Hannah Montana. Didn't work.

The coolness thing is that now, a few years after that conversation, when I mention Hannah Montana, Nieves raises an eyebrow and looks at me almost confused and asks, "Who?"

I, sometimes, want to be Tony Robbins! Okay, I don't really want to be him. I've never met him, although I would be honored to. I want to be able to do what he does, and that's to inspire people to motivate themselves. And make money at it! Okay, the money really isn't an incentive for me. Really! What I really want to do is help others empower themselves to be their best. Guess what?! I DID that! For decades! Actually, I'm still doing it.

My point here is that Cheryl IS right! But getting there is long! Or hard. We have to mature to that idea and, more often than not, it takes some experience. Often, HARD experiences.

INSIGHT: Here's what I learned from this conversation with my friends and within myself. Sometimes hipness is what it ain't! It's not about being someone else, in someone else's skin. It's about being YOU! Dig deep. Find you. And let you shine! THAT's hip!

If you could, how would you?

I walk into the boss's office one day only to find his exec about in tears, he was so frustrated. I ask him, "Are you okay, Sir?" "No, Chief, I'm not," he replies. "What's up and how may I help you?" I ask him. "I'm incredibly upset at those folks in the finance department. I'm trying to get them to pay for a guest we've invited to speak to the team. We have the money. It was designated for exactly this function. They just keep telling me that they can't do it," he tells me.

"Sir, maybe you need to ask from a different angle, ask a different question of them," I suggest. "I'll try anything, Chief. What do you suggest?" he asks. "Call them back and ask them this: 'If you COULD do it, how would you?' When they give you the answer, ask them to do it then." His eyes still watery from his previous conversation with the folks in the finance department, he almost smiles and says, "I'll try it. The boss will see you now." I walk in to see the boss.

As I come out of the boss's office, I notice his exec smiling. He comes up to me and shakes my hand. "Thank you, Chief! I called them back and asked them your question. They told me exactly how they could make it happen. I "asked" them to do it just as they told me. They made it happen! Thank you!"

INSIGHT: Sometimes, we have to ask a different question to get people to think differently. Especially when the answer to the first question is no. There's always a way. But the path to that way may be obstructed by old paradigms and methods. If you could, how would you?

What's important?

Man, I was so proud of myself! It was about noon when I got home and as I walked in the house, I found my lovely-bride-of-16-wonderful-and-fulfilling-years, Deb, and my 10-year-old daughter, Tesa, having a discussion. Okay, they were arguing. Oh, all right, they were fighting! I didn't know, nor did I really care, what they were fussing about. I had a very important announcement to make! I quieted them down and proudly announced that I had just been selected to be the Commandant of the March Air Force Base Noncommissioned Officer Academy. Hey, this was such an historic event I left work two hours early! I gave up my daily tee time to go home to share my excitement with my family! I'd worked so hard to reach that pinnacle of success! No other bandsman had ever done that! "Man, I'm good!" I thought. "I've made it to the top!"

As I expected, I could see the pride in Deb's green eyes. Tesa, our first-born, looked at me as intently as I'd ever seen her look, and without a moment's hesitation asked, "Fine, Dad, now will you help me pick up poop?" Oh, as you may imagine, my ego immediately deflated to almost nothing.

You see, although my career was of the highest importance to me, what was much more important to Tesa was what I often considered mundane minutiae. It's the little things that our children and our families find important. Often, our children won't know, nor care, what we do as a profession, but they'll always know whether we were there when they needed us. And they will always remember. Tesa still remembers, decades later.

I did go out and help Tesa pick up poop. Afterward we sat down on the stoop and she said to me, without any prompting, "Dad, I'm proud of you." I never asked her if she was referring to my news or my help. It didn't matter. She was proud of me and that's all that mattered that day.

INSIGHT: Here's the lesson. Pay attention to what your family needs and be there for them. It may be difficult, but just say no to work and yes to family more than you have in the past. It doesn't matter how often you've already done it, it hasn't been enough, trust me.

With respect to....

I worked with a person, Mary, for several years right after I retired from active duty. We became friends and colleagues. She was still on active duty...a lieutenant colonel. She retired about two years after we'd started working together. While she was on active duty, how I referred to her was simple. I called her "colonel." I'd grown to know her well enough to know that she wasn't going to accept me calling her "colonel" after she retired, so I thought and pondered, "what shall I call her?" I wanted to maintain my respect for her. She thought more liberally than I, so she told me to call her Mary. NO WAY! I'm from the South! South New Mexico! But even there, I was taught to call people Mister or Missus, or Mizz, but NEVER by their first name. So, I called her Mizz Mary. All with the intent of being respectful. We worked together another five or six years, all the while with me calling her Mizz Mary.

One day we were chatting, just shooting the breeze, when the topic of how we address people came up. I was freaked out when she told me that she thought people calling her “Mizz,” as in Mizz Mary, was disrespectful! I’d had to work hard to come up with a respectful form of address for her. And I’d had to practice to do so whenever I saw her. Now, she tells me, she thinks it’s disrespectful?! I never addressed her again. I didn’t know how. We still worked together for a year or two before she retired again, and I was sure not to disrespect her.

INSIGHT: I decided, after that obviously silly mistake on my part, to do what would, surely, be respectful. I’ve since then, asked. Asking someone how they’d like to be addressed is the best way to ensure you’re respectful. Now, I still insist on putting a Mister or Missus in front, and I tell folks who expect me to call them by their first name, that I feel respected when people call me “Chief.” I feel uncomfortable calling people who call me Chief by their first name, even when they’re a lot younger than I. We don’t always know what is respectful to others. Just ask….

Information is NOT power!

I walk into my home and see my lovely bride, Deb, smiling. "Hey, Baby," she says. "Hey, Baby," I reply, "How are you?" We go on, asking the usual questions about each other's day, how it went, what we did, etc. Then we get to a critical part of the conversation.

"You know," Deb says, "I was talking with Lori today." Lori was married to one of my colleagues. "Oh?" I asked, "what about?" "Well, we were just chatting when she mentioned something about you guys being gone for Thanksgiving."

I was on active duty when this occurred. I was assigned to the United States Air Forces in Europe Band at the time, stationed in Germany. One of the coolest things we did in that band, and I take at least partial credit for creating the concept, was that we deployed a big show we called Operation Seasons' Greetings to locations where our troops were. We had great success with those shows because we brought in artists from the United States and did a big production for our troops and their families, similar to a USO show. I'd had this great idea of playing for the troops (no families there) stationed in Bosnia for Thanksgiving Day! I'm sure you're saying, "What a GREAT idea!" Everyone I'd talked with said the same thing, "What a GREAT idea!" With one exception....

"Lori told you we'd be gone for Thanksgiving?" I asked Deb, trying to sound as if...okay, the OC Factor kicked in. "Oh, Crap!" was all I could think of. In my zeal to get the show together and make it a great event for the troops, I'd forgotten, no kidding, I'd just forgotten, to tell the most important person in my life what I was up to and that it would affect her and the girls since I wouldn't be home for Thanksgiving.

Here's the point. Remember that old saying, "information is power?" That's stupid! Information is just information. It won't get up and do anything powerful! It'll just sit there, sometimes in your brain, and do absolutely nothing! Imagine being the smartest person in the world and no one knowing it? What good would that be? SHARED information,

APPLIED information is power.

INSIGHT: If you know something that someone else should need to know or needs to know to make their life better, SHARE IT! That's how you become powerful! Deb forgave me. I, of course, bought her a new BMW....

Write!

For about three decades I was a writer. I wrote music. I painted pictures with sounds. That was before I started painting pictures with words. Both have been seemingly right for me.

The first mentoring I ever received about writing music was from an incredible writer and leader who eventually became a friend and mentor. He was Senior Master Sergeant Norm Charpentier. He was the “typical” musician. Great player and leader. Trombonist. He had stories for most topics and used many descriptive words when sharing them. He was a great music arranger/writer.

I quickly became his student. His first and best, still powerful, lesson looked something like this:

Me: Norm, I want to be a writer. A good one, like you. What do I need to do?

Norm: Write.

Me: Yes, that's what I want to do. How do I get to be good.

Norm: Write.

Me (getting frustrated): Yes! That's what I'm asking. How do I become a good writer?

Norm (frustrated as well) as he walks away: Write!

You see, all this time I thought Norm was saying "RIGHT!" Not so. He was answering my question. There are three lessons here. First, ensure you're communicating by using language that your listener is sure to understand. I'm not blaming Norm. He was just in a succinct mood during that conversation. Secondly, listen for meaning, not just words. I thought I was hearing what I was hearing but it turned out I was hearing something different than what Norm was saying. Sounded the same to me. Thirdly, if you want to get good at something, it takes practice. Do it! Eventually, you'll get better, or at least good enough to seek further training. I did. I wrote a lot of music for about four decades. I won't say I became good at it, that might seem egotistic of me. The folks who played my music would have to decide that, along with those who heard it. But I stayed employed doing it, so I must have been, at least, decent.

INSIGHT: Whatever it is you aspire to do, even become. Don't be afraid to try. Do your best at it. Seek help and keep trying. RIGHT?!

Life-long service…

As I write this, I just went over forty-eight years of service to my country. I joined the Air Force on 1 February 1972, you see.

A few years ago, I was leading a character workshop for my freshmen at the United States Air Force Academy that my second-born daughter, Elyse, was observing. That workshop always started with the Permanent Party Facilitators introducing themselves so that the freshmen could see who was in their presence. I'd have the Facilitators tell the cadets who they were, where they were from, what they did in their service, how long they'd served, and so on.

Elyse was the last to introduce herself. As she stood up, she mentioned that she was my daughter and that she'd served in the Air Force all of her life. WHAT?! I'd never thought of that! She was born on an Air Force Base, at Keesler Air Force Base in Mississippi, and traveled most of the United States and a lot of Europe as what we call a “dependent.” Freaked me out! She was exactly right! She’d served her entire life!

INSIGHT: Now, as unique as my baby is, and she is, trust me, she’s not different than any other of the hundreds of thousands of children who have served in the Air Force and probably haven’t ever thought of it. That’s service! Let’s not forget their contributions. We wouldn’t be us without them!

Closing thoughts...

Well, I've run out of wisdom to share with you. As I said in the introduction, I hope that these stories resonate with you and that you may even relate to some of them. Years ago, I adopted a War Cry that encapsulates my purpose in publishing these stories, that's HEIRPOWER!

One afternoon, while serving as the Deputy Director for the Family Support Center at Ramstein Air Base in Germany, I'm sitting at my desk, just thinking...pondering nothing in particular, when a thought, an insight, comes to me, serendipitously, of course. We, Americans in particular, are SO blessed, yet we often take our blessings for granted. Mostly because we "get caught up in the thick of thin things," as Dr Stephen Covey would admonish us. These stories, the thoughts generated by them, and the Wisdom Insights I've shared with you here, are the fruit of paying attention and being grateful for what's happened around me for the past fifty-or-so years. There are more blessings and more stories that I'll eventually capture in volume two. I'm going to keep paying attention and appreciating the many blessings that the Creator bestows upon me and my loved ones. That, I think, is the essence of wisdom. And the effect of the insight will be in sharing it with others, you, in particular. That's HEIRPOWER!

You see, the concept of HEIRPOWER! is simple. An heir is a successor. If you break that word apart, you're left with "success" and "or." Success or.... The alternative to success is failure. HEIPOWER! is about enabling people to empower themselves to succeed. Power is the capacity to act effectively. HEIPOWER! is produced when we share what we have. I don't have a LOT of wisdom, but what I do have I'll always, gladly, share with you.

Until next time, I wish you all the best and all the wisdom you need to have a rich and fruitful life.

HEIRPOWER!

bob vásquez!

Other books by bob vásquez available at amazon.com…

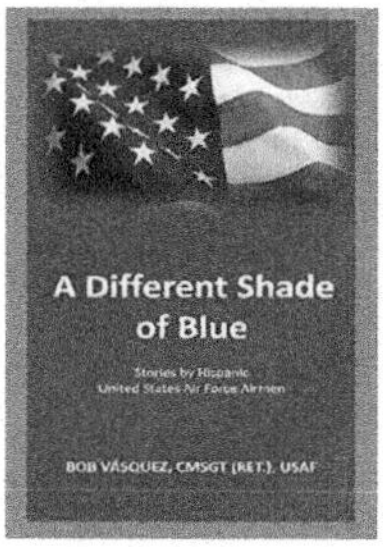

Made in the USA
Monee, IL
04 November 2026

67375692R10069